ADAM
The Missing Link
(The New History of Mankind's Creation)

By Marshall Klarfeld

Trafford
PUBLISHING®

Note for Librarians: A cataloguing record for this book is available from Library
and Archives Canada at www.collectionscanada.ca/amicus/index-e.html

Printed in Victoria, BC, Canada.

ISBN: 978-1-4251-9184-9

*We at Trafford believe that it is the responsibility of us all, as both individuals
and corporations, to make choices that are environmentally and socially sound.
You, in turn, are supporting this responsible conduct each time you purchase a
Trafford book, or make use of our publishing services. To find out how you are
helping, please visit www.trafford.com/responsiblepublishing.html*

*Our mission is to efficiently provide the world's finest, most comprehensive
book publishing service, enabling every author to experience success.
To find out how to publish your book, your way, and have it available
worldwide, visit us online at www.trafford.com/10510*

www.trafford.com

North America & international
toll-free: 1 888 232 4444 (USA & Canada)
phone: 250 383 6864 • fax: 250 383 6804
email: info@trafford.com

The United Kingdom & Europe
phone: +44 (0)1865 487 395 • local rate: 0845 230 9601
facsimile: +44 (0)1865 481 507 • email: info.uk@trafford.com

10 9 8 7 6 5 4 3 2 1

Contents

Chapters

Index of Photos and Graphics
The Irrefutable Physical Evidence

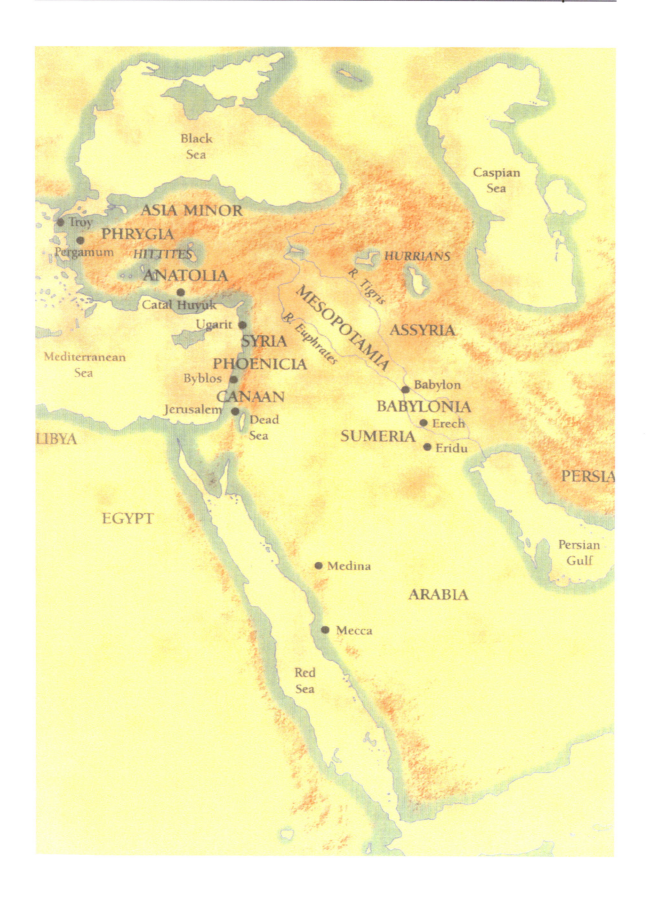

As an undergraduate student at CALTECH in the late 1940's, I was fascinated by the advanced scientific knowledge found in the Bible's story of creation. Genesis describes the creation of our solar system as occurring in a 6 "day" time frame.

01:016 "And God made two great lights; <u>the greater light to rule the day</u>,
    and the lesser light to rule the night; he made the stars also."
01:017 "And God set them in the firmament of the heaven to give light upon
    the earth,"
01:018 "And to rule over the day and over the night, and to divide the light
    from the darkness: and God saw that it was good.
01:019 "And the evening and the morning were <u>the fourth day</u>".

Before the ignition of our Sun, there could not be any "days". Scientific research has revealed <u>that this ignition event</u> always occurs **late** in a solar system formation process. My question was, how did this 5000 year old Bible description of the birth of our solar system disclose the advanced scientific information (the ignition of our Sun occurs at the 4th "day", in a 6 "day" creation process), thousands of years before we discovered it?

Uncertain how to unravel this puzzle by myself, I questioned my Nobel laureate professors, Linus Pauling and Richard Feynman. Their powerful answers stayed with me for decades and inspired me to pursue the knowledge that I am now able to share with the readers of this book.

My enduring quest to solve "the puzzle" finally led me to "The Earth Chronicles", a series of books written by Zecharia Sitchin. His meticulous translations of thousands of cuneiform tablets, as presented in his nine books and numerous articles, provided me with answers that solved "the puzzle" presented by Genesis.

In my conversations with Sitchin, he stated, "It is not a secret. I am just a reporter." I believe his translations of the early knowledge of our solar system and the origins of humanity, will be recognized one day as among the most significant historical findings ever. Zecharia Sitchin's astonishing accomplishment, in my opinion, is deserving of the Nobel Prize.

Human civilization's first records were discovered on thousands of clay tablets, unearthed in the ruins of the "cradle of civilization", an area called Mesopotamia in the early 1800's. The tablets were inscribed with the first written language called *cuneiform text* by anthropologists. Clay tablets have survived for thousands of years because they were preserved by the heat of fires invading armies instigated to destroy conquered Mesopotamian territory.

The information revealed on these cuneiform tablets is the history of earliest human civilization and its roots in the world of the incomprehensibly intelligent extraterrestrial species who visited planet Earth 450,000 years ago. They called themselves *Anunnaki*.

Landing in the body of water now called the Persian Gulf, the *Anunnaki* established their first settlement at ERIDU (home in the far away). Archeologists have uncovered the sites of five other settlements, located in what is now southern Iraq, that predate the first human civilization of Sumer.

The Sumerian civilization produced so many firsts that experts are as yet unable to explain how these early humans invented writing, mathematics, astronomy, astrology, the wheel, the plow, and religion. The Sumerian language is totally unrelated to any other known language. The information revealed on the cuneiform tablets clearly states that it was the *Anunnaki* that <u>gave</u> the Sumerians "everything."

The Epic of Enuma Elish (The Story of Creation) was found written on seven clay tablets first discovered by Henry Layard, in the ruins of the library of Ashurbanipal in Nineveh. This entire story was recited at every New Year's ceremony for thousands of years before it was preserved on clay tablets. (Zecharia Sitchins's book "When Time Began" page 4---"One long text, written on seven tablets, has reached us primarily in its later Babylonian version. Called the *Epic of Creation* and known by its opening words *Enuma elish*, it was publicly read during the New Year festival that started on the first day of the month of Nissan, coinciding with the first day of spring.")

The Epic of Gilgamesh was found written on twelve clay tablets and it's recognized as Humanity's first recorded story. It is entirely probable that this historically significant story also was transmitted orally for many generations before being recorded on clay tablets. Evidence suggests that this important story was given to the Sumerians by the Anunnaki. The express purpose of this story was to transfer important facts to future generations to help unravel the numerous mysteries about the survival of humanity during the Great Flood. The concept of destiny and fate is an essential part of this story.

Cuneiform is humanity's oldest writing system. The British Museum contains over 120,000 cuneiform tablets. The libraries at Nineveh & Nippur were discovered in 1845 with scattered tablets (shown here) from collapsed shelving. In excess of 300,000 cuneiform tablets have been found to date.

The most famous cuneiform tablet is The Great Flood story (shown at center page). It is the 11th tablet of the 12 tablets called Epic of Gilgamesh.

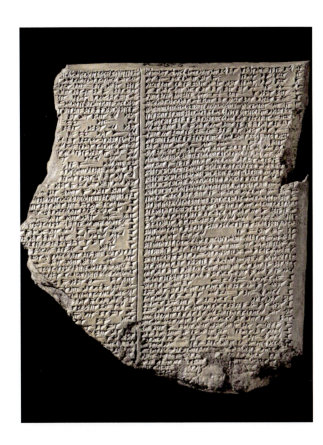

Cuneiform writing was accomplished with smooth wet clay and a wedge-shaped stylus. The Latin word cuneus means "wedge." Written language originated in Sumer over 6,000 years ago. This form of communication was adopted by the Akkadians, Babylonians and Hittite's and was used for over 3,000 years. Modern technology is searching for a storage scheme that will last as long as these tablets.

Modern translations have been aided by 3D model scanning and special non-photorealistic shading.

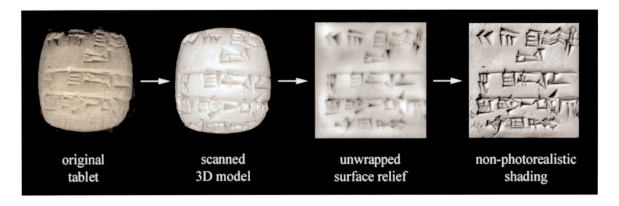

| original tablet | scanned 3D model | unwrapped surface relief | non-photorealistic shading |

This book contains research which I have divided into four categories to help explain "technology transfers." These transfers were left to humanity, by the *Anunnaki*, to help supply answers to these questions:

"How was our solar system shaped?"

"How did humanity evolve on our planet?"

The four categories are:

1. <u>S.I.T.</u>   <u>S</u>tored <u>I</u>nformation <u>T</u>ext – Cuneiform tablets

2. <u>S.I.D.</u>   <u>S</u>tored <u>I</u>nformation <u>D</u>evice – Cylinder seals

3. <u>S.I.Sc.</u> <u>S</u>tored <u>I</u>nformation <u>Sc</u>ulpture – Ishtar statue, etc.

4. <u>S.I.St.</u> <u>S</u>tored <u>I</u>nformation <u>St</u>ructure – Pyramids, etc.

In the recorded history of planet earth, it has been observed that when an "advanced civilization" encounters a primitive culture, several things tend to occur.
- The advanced civilization dominates the primitive culture.
- The advanced civilization imposes its "form of governing."
- The advanced civilization "transfers its technology."

This book illustrates the long-term preservation of certain important "technology transfers" from the *Anunnaki* civilization to the early civilizations of planet Earth.

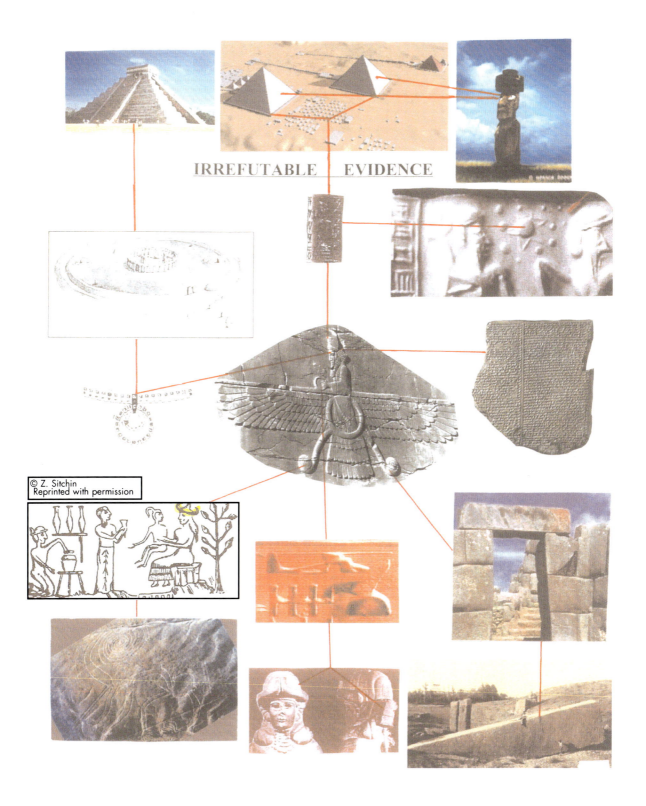

IRREFUTABLE EVIDENCE

© Z. Sitchin
Reprinted with permission

GILGAMESH	Headstrong King seeking immortality
ISHTAR	Beautiful seductress, Princess of Nibiru
NIBIRU	10th Planet of our solar system
ANUNNAKI	Inhabitants of Nibiru
ANU	Ruler of the Anunnaki
ENKIDU	Cloned friend of GILGAMESH
ZIUSUDRA	Hero of the Deluge, Bible's NOAH
ENLIL	Commander, builds landing platform
ENKI	Brilliant genetic engineer, creator of Adam
THOTH	Son of Enki, builder of Pyramids
BULL OF HEAVEN	Powerful laser-like trencher

Cylinder seal of GILGAMESH and ENKIDU, the DNA clone of the King. Note the double helix column both are grasping.

ISHTAR, the powerful Goddess of Love and War. She stands atop a Lion, with weapons on her back and a bow in her left hand.

Tablet #6 tells of the beautiful seductive Princess Ishtar. Seeing the handsome King of Uruk, she lusts for his love. She leaves "above" (the orbiting Mother ship) and descends (via shuttle craft and "whirlwind") to meet with Gilgamesh and make her proposal.

Gilgamesh rudely refuses. This refusal not only angers Ishtar, it infuriates her. She returns to the Mother ship and threatens Anu (her grandfather and leader of her clan) in order to gain possession of the Anunnaki's most powerful machine—"The Bull of Heaven." Taking this laser-like device back down to Uruk, she begins to blast huge 200-foot deep craters in the streets, killing hundreds of people. Gilgamesh and Enkidu attack "The Bull" and destroy it, hurling part of it at the retreating Ishtar.

In this 6th tablet we are told about one of history's most incredible women- Ishtar. She is described as the "Goddess of Love." We learn that she has the favor of Anu (described by the Sumerians as the leader of the Anunnaki – "those who gave everything to make life possible".) We also see that Ishtar is able to travel from Anu's place down to earth and back again. Armed with the powerful "trencher," she gives vent to her rage. Unable to punish Gilgamesh, because he has 2/3 royal blood, she causes Enkidu to die.

Pg. 5. A 4000 year old life-size statue of Ishtar shows her remarkable beauty and her space helmet and suit. (S.I.Sc.=Stored Information Sculpture)

Pg. 7. "Ancient Egyptian Flying Vehicles." (S.I.Sc.=Stored Information Sculpture)

Tablet #11 is the fascinating story of the great flood. Gilgamesh's adventures led him to his great Uncle Ziusudra who tells him the secret story hidden from humans, of how he was chosen to save humanity. The story encompasses the advanced knowledge that the Antarctic ice shelf would slide into the ocean producing an enormous tidal wave. The secret meeting of the council of the 12 Great Lords of the Anunnaki in the course of which Enlil obtains their sworn oath "not to save anyone." Enki's dream message from the "Creator of All," describes the plan to have Enki tell the wall of his son's home (Ziusudra) how to build the water tight vessel that will save humanity. After the water subsides and the vessel is grounded, the message of this Epic is revealed. Furious that humanity has survived, Enlil confronts Enki and accuses him of breaking his oath. Enki describes his dream and declares that it was destiny that brought them to this planet to create humanity, and that it is destiny that humans are to inherit planet Earth. Enlil is persuaded that if the "Creator of All" has revealed their destiny, then they must help humanity to survive. Enlil relents and bestows long life to Ziusudra and his wife. The brothers lock arms in a sign of unity.

(Earth King's Friend) ENKIDU
Adult clone with Gilgamesh's DNA

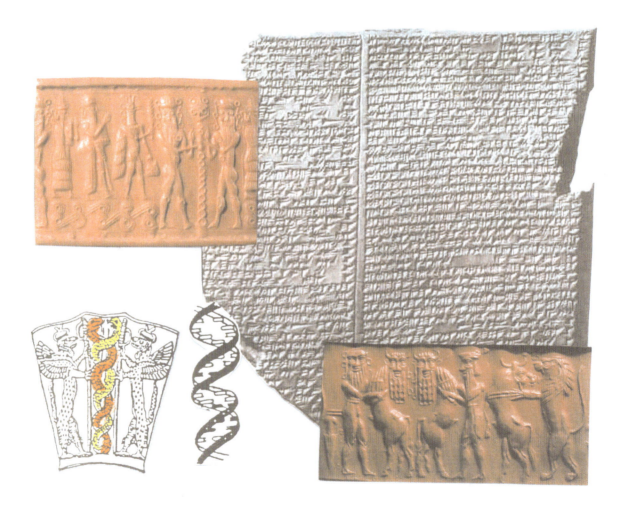

(Earth King) Gilgamesh
The First Action Hero
His mother is an Anunnaki Princess and his father is human
(2/3 ANUNNAKI, 1/3 human)

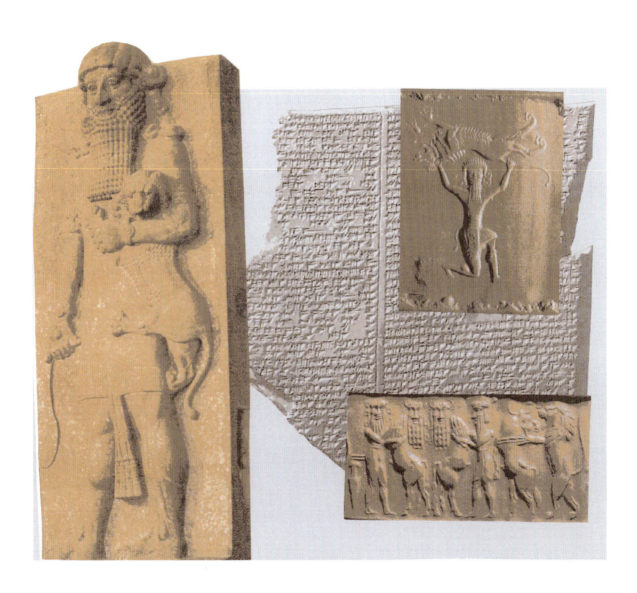

Epic of Gilgamesh

There are many translations of the 12 cuneiform tablet Epic of Gilgamesh. The work contains some 3,000 lines of text and is considered history's <u>first</u> written story. In my view this story is a message left to future generations, by those who knew of the events described in the text, and wanted to transfer this information. (<u>S.I.T.</u>=<u>S</u>tored <u>I</u>nformation <u>T</u>ext)

Pg. vii. Gilgamesh and the characters that appear in this story.

To the casual reader, Gilgamesh is the story of a King of Uruk, whose mother is a Divine Princess and his friend, Enkidu, who is suddenly stricken and dies. Fearful of death, Gilgamesh seeks out his uncle (survivor of the Great Flood) to achieve immortality.

To me, Gilgamesh is much more than just the above story. It includes a procreation relationship between a god and a human. In Genesis 6:04:

> "The Nephilim were on the earth in those days, and also afterward,
> when the sons of God came in to the daughters of men, and they bore
> children to them. These were the mighty men that were of old, the
> men of renown."

Gilgamesh's mother was a Divine Princess and his father was human. Gilgamesh is 2/3 Divine and thus feels he could become immortal.

If we look factually at Gilgamesh's friend, Enkidu, we see that the Anunnaki sent a cloned man to help the people of Uruk. Enkidu is described as a wild hairy man with enormous strength. He is civilized by a local harlot and becomes best friends with Gilgamesh. Initially they fight, with equal strength, but as they look into each other eyes, they see themselves. (Note this DNA match of Gilgamesh is named after Lord Enki, the Anunnaki's chief genetic engineer.)

In 2004 we are currently on the threshold of cloning babies, yet in this 6,000 year old story we are told of the ability to clone a material match of a grown man.

(Space Princess) ISHTAR
ANUNNAKI Goddess of Love and War

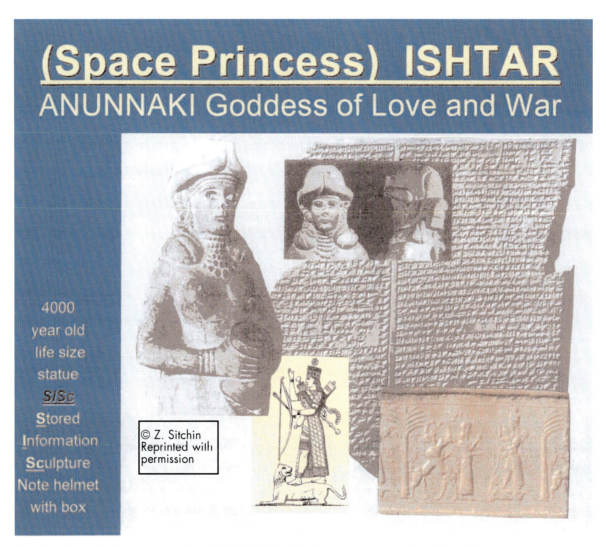

On her head she wears the SHU.GAR.RA helmet – a term that literally means "that which makes go far into universe."

It was from this historical juncture that the Anunnaki helped advance humanity through incredible transfers of technology and agriculture to the Sumerians.

Historians are still unable to explain the sudden arrival of "The Sumerian Civilization." It produced so many "firsts"...writing, mathematics, Law, astronomy, astrology, the wheel, the plow, and religion. The Sumerians built many cities with significant architectural buildings and temples.

Pg.8. Enlil giving the plow, Enki bringing animal husbandry, astronomical
map of our complete solar system showing their planet Nibiru.
Cylinder seals, like VA/243 were precision crafted information
storage devices made from stone. (S.I.D.=Stored Information Device)

The intent of the images on seal VA/243 is the equivalent to Pioneer 10's plaque and to the Voyager 1 & 2 GOLDEN RECORDS.

 Pg.9. NASA sent these images out into space to show "who we are, where the space craft came from, and the location of our star in our galaxy." (S.I.D.)

The story of Gilgamesh contains Stored Information that informs us of:

1. The Anunnaki cloned an adult man from the DNA of the King.
2. There was interface between earthlings and the Anunnaki as shown when the people of Uruk plea for relief from their King's bad behavior, and the Anunnaki create Enkidu.
3. The dramatic story of the Great Flood is told; the message implied is that the destiny of Earthlings, created by the Anunnaki, is to inherit the Earth.
4. Ishtar, with her space helmet and shuttlecraft, is able to travel back and forth between Earth and the Mother ship.
5. The "Bull of Heaven," a powerful laser device, is capable of trenching and leveling mountain tops.

ANCIENT FLYING MACHINES AND "THE ARK"

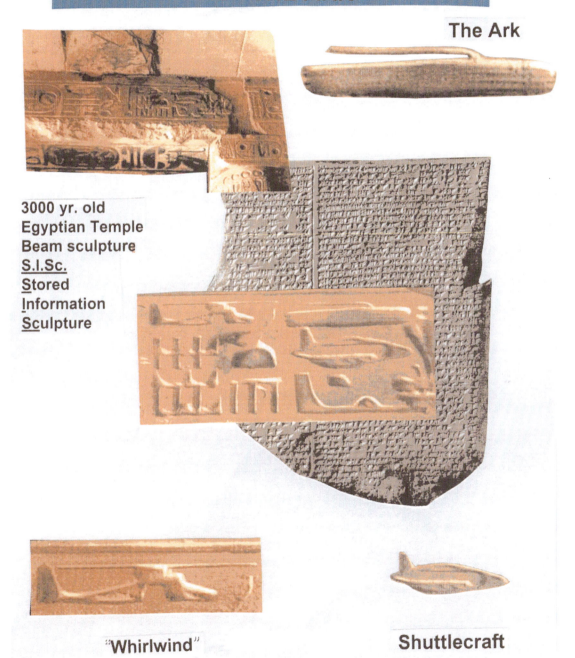

ANCIENT FLYING MACHINES AND "THE ARK"

The Ark

3000 yr. old
Egyptian Temple
Beam sculpture
S.I.Sc.
Stored
Information
Sculpture

"Whirlwind"

Shuttlecraft

Sun in the Center - Our Solar System

4,500 year old tablet - 2,500 B.C.

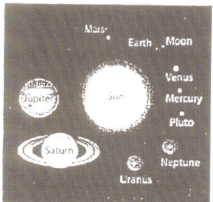

© Z. Sitchin
Reprinted with permission

Cylinder seal VA/243 is approximately 4,500 years old. When this small (2.7 cm tall) stone cylinder is rolled across wet clay (see below), a positive picture story is created. VA/243 shows two Anunnaki leaders (horned headdress) in extraordinary detail. Seated (ENLIL) is gifting the plow, and (ENKI) is bringing animal husbandry. The diagram of our solar system with 10 planets is presented in remarkable scale. The 10th planet (NIBIRU) is the Anunnaki's home planet. Creating this very small and beautifully detailed negative image on this very hard polished hematite cylinder, demonstrates an extremely advanced technology. The engraved diagram of our solar system reveals the excellent astronomical knowledge of the creators of this S.I.D. (Stored Information Device).

Cylinder Seal VA/243

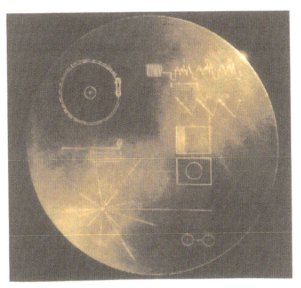

VOYAGER 1 & 2 carried Carl Sagan's Golden Record with spoken greetings from Earth-people in fifty-five languages. The first spoken language was Sumerian.

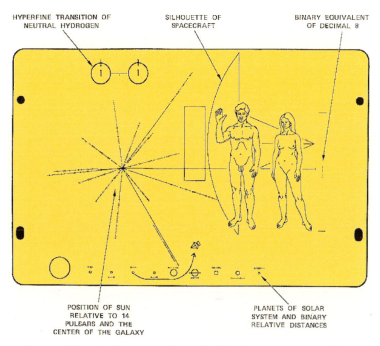

PIONEER 10 PLAQUE, designed by Carl Sagan, shows Earth's location in our solar system, position of the sun in our galaxy, and our species (Homo sapaien–ADAM & EVE).

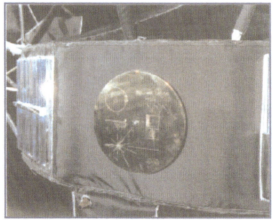

VOYAGER'S INTERSTELLAR OUTREACH PROGRAM

VOYAGER'S GREETINGS TO THE UNIVERSE

The voyager spacecraft will be the third and fourth human artifacts to escape from the solar system. Pioneers 10 and 11, which proceeded Voyager in out-stripping the gravitational attraction of the Sun, both carried small metal plaques identifying their time and place of origin for the benefit of any other spacefarers that might find them in the distant future. Using this example, NASA produced a more ambitious message aboard Voyager 1 and 2. The Voyager message is carried by a phonograph record (a 12 inch gold-plated copper disk that contains sounds and images portraying the diversity of life and culture on Earth.) A NASA committee chaired by Carl Sagan produced 115 images and many natural sounds such as the surf, the wind, thunder, birds, whales, and other animals. To this they added musical selections from different cultures and eras, and spoken greetings from Earth-people in fifty-five languages.

Carl Sagan

Carl Sagan – described by Elz Cuya as "a scientist, philosopher, and poet – popularized science, an otherwise cold discipline reserved for the academic world. And with his eloquence and humanity, Sagan brought science to the foreground for all of us to poke at, question and marvel."

"His 1980's television series *Cosmos* discussed other galaxies, <u>challenged our notions of planetary systems, and probed the origins of life</u>. But unlike most scientists, Sagan didn't dismiss the topics of science that captivated people's imaginations and fantasies such as astrology, UFO's, the possibility of extraterrestrial life and the theories of space travel."

Elz Cuya went on to state, "Yet Sagan never took the elitist stance as a disciple of the scientific world. He humanized it, making it less sterile. He renamed the scientific method, 'The Baloney Detection Kit.' He admitted that science at times had been wrong. However, if a hypothesis is wrong, "it provides a structure to find out what's right." Indeed, Sagan had the ability to enlighten the world while at the same time, humbling us in our humanity. Sagan said, "Somewhere, something incredible is waiting to be known."

The back cover of Carl Sagan's 1985 book "Contact" shows the author leaning against a building with the "Winged Planet" design engraved on this structure above his left shoulder. (pg. 10) Was this a deliberate attempt by Sagan to connect his story of first contact with the Sumerian stories of the Anunnaki, an extra-terrestrial race? He names a spaceship Gilgamesh.

When coupled with Dr. Sagan's input on the Voyager 1 & 2 GOLDEN RECORD messages – greetings in 55 languages, the first being Sumerian, a 6,000 year old language that is no longer spoken. I suspect that Dr. Sagan was a believer who wanted to signal his beliefs, without overtly confronting the scientific community.

ENUMA ELISH

The Epic called Enuma Elish ("When on High…") was first discovered by Henry Layard, in the ruins of the library of Ashurbanipal in Nineveh. It was written in cuneiform text on seven tablets, each containing between 115 and 170 lines. There are many translations of this Epic (E.A. Speiser Princeton, 1969, and L.W. King, The Seven Tablets of Creation, London 1902.)

George Smith was the first to publish his translations in 1876, under the title The Chaldean Genesis. The discovered language was Akkadian written in an old Babylonian dialect. An earlier version written in Sumerian suggests that the Epic traces its origin to the Sumerian civilization. This civilization is credited with inventing cuneiform writing and thus it might be assumed that Enuma Elish dates back 6,000 years or earlier.

As with almost all ancient stories, years of oral transfer occurred before writing was invented. It is also my assumption that Enuma Elish was given to the Sumerians, as a creation story, by the Anunnaki. The fact that there were seven tablets and many parallels to the accounts in Genesis, caused some scholars to suggest that the Genesis narratives were rewrites of the Sumerian story. (The "god" finished his creation work within the span of six tablets. On the last and 7th tablet, "god" exalted in the handiwork and greatness of his creation.)

Pg.13. displays a translation from the first tablet. The story of how our solar system came to its current configuration is traceable to the events described in this Epic. As interpreted by Zacheria Sitchin (Pg.13) the names of the "gods" are the names of our solar system's sun, moon, nine planets, and the 10th planet (Nibiru). Nibiru enters the system from a clockwise direction, encountering the watery planet Tiamat, and Nibiru's moon (Northwind) blasts a hole in Tiamat. This collision forced Tiamat into a new orbit and is currently the planet Earth. The excess debris from this collision formed the Asteroid Belt (The Hammered Bracelet) and the comets orbiting in the same clockwise direction as Nibiru. According to Sitchin's translation, Nibiru was captured by Neptune's gravity and pulled into its current 3,600-year orbit around our Sun. This celestial collision could have been the contact that seeded planet earth with life-producing DNA.

Enuma Elish:

When on high the heaven had not been named,
Firm ground below had not been called by name,
When primordial Apsu, their begetter,
And Mummu-Tiamat, she who bore them all,
Their waters mingled as a single body,
No reed hut had sprung forth, no marshland had appeared,
None of the gods had been brought into being,
And none bore a name, and no destinies determined--
Then it was that the gods were formed in the midst of heaven.
Lahmu and Lahamu were brought forth, by name they were called. (10)

Before they had grown in age and stature,
Anshar and Kishar were formed, surpassing the others.
Long were the days, then there came forth.....
Anu was their heir, of his fathers the rival;
Yes, Anshar's first-born, Anu, was his equal.
Anu begot in his image Nudimmud.
This Nudimmud was of his fathers the master;
Of broad wisdom, understanding, mighty in strength,
Mightier by far than his grandfather, Anshar.
He had no rival among the gods, his brothers. (20)

Thus were established and were... the great gods.
They disturbed Tiamat as they surged back and forth,
Yes, they troubled the mood of Tiamat
By their hilarity in the Abode of Heaven.
Apsu could not lessen their clamor
And Tiamat was speechless at their ways.
Their doings were loathsome unto . . . .
Their way was evil; they were overbearing.
Then Apsu, the begetter of the great gods,
Cried out, addressing Mummu, his minister: (30)

"O Mummu, my vizier, who rejoices my spirit,
Come here and let us go to Tiamat!"
They went and sat down before Tiamat,
Exchanging counsel about the gods, their first-born.
Apsu, opening his mouth,
Said to resplendent Tiamat:
"Their ways are truly loathsome to me.
By day I find no relief, nor repose by night.
I will destroy, I will wreck their ways,
That quiet may be restored. Let us have rest!" (40)

As soon as Tiamat heard this,
She was furious and called out to her husband.
She cried out aggrieved, as she raged all alone,
She uttered a curse, and unto Apsu she spoke:
"What? Should we destroy that which we have built?
Their ways indeed are most troublesome, but let us attend kindly!"
Then Mummu answered, giving counsel to Apsu;
Ill-wishing and ungracious was Mummu's advice:
"Do destroy, my father, the mutinous ways.
Then you will have relief by day and rest by night!" (50)

When Apsu heard this, his face grew radiant
Because of the evil he planned against the gods, his sons.
As for Mummu, he embraced him by the neck
As that one sat down on his knees to kiss him.
Now whatever they had plotted between them,
Was repeated unto the gods, their first-born.
When the gods heard this, they were astir,
Then lapsed into silence and remained speechless.
Surpassing in wisdom, accomplished, resourceful,
Ea, the all-wise, saw through their scheme.

Apsu	= Sun
Mummu	= Mercury
Lahamu	= Venus
Lahmu	= Mars
Tiamat	= Earth (after)
Kishar	= Jupiter
Anshar	= Saturn
Anu	= Uranus
Ea	= Neptune

© Z. Sitchin
Reprinted with permission

EPIC OF CREATION (ENUMA ELISH)

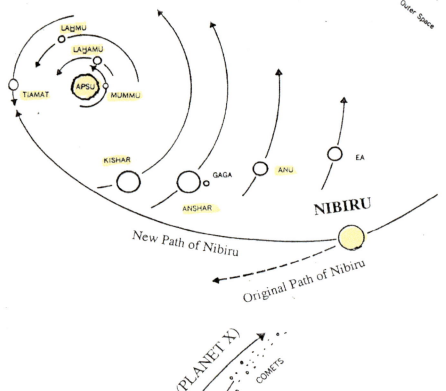

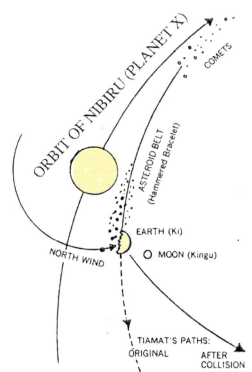

My instinct, when I first read this story of the formation of planet Earth and our moon (Kingu), was to search for facts that might support these events. The translations of ENUMA ELISH by Zacheria Sitchen, falls into my S.I.T. (Stored Information Text) category. If this was indeed a transfer of facts, as I assumed the story of the Great Flood in GILGAMESH was, then I needed additional evidence that planet Earth is a sphere with a huge hole in it.

Pg. 16 shows the current form of planet Earth with the waters of the Pacific Ocean (33% of the global surface) occupying the remains of the hole that was blasted out of planet Earth by the events described in Enuma Elish. The depth of the Mariana Trench is 6.5 miles below sea level. Mount Everest could stand in this hole and one mile of water would cover its peak.

Pg. 17 illustrates that continental drift is causing the land mass called Pangea to redistribute its mass as a result of an unbalanced condition. 225 million years ago the land mass called Pangea contained today's six continents, clustered together on one side of the planet. If there was not a similar land mass on the other side of the planet to balance the rotating sphere, then an "out of balance" condition would exist. Thus the "out of balance" forces being exerted today are still working to correct for the loss of mass caused by the impact of Nibiru's moon (Northwind). It is called "continental drift."

After these revelations of how our solar system came to its current condition, Enuma Elish continues with the story of the creation of humanity. This story will be discussed in Chapter 6.

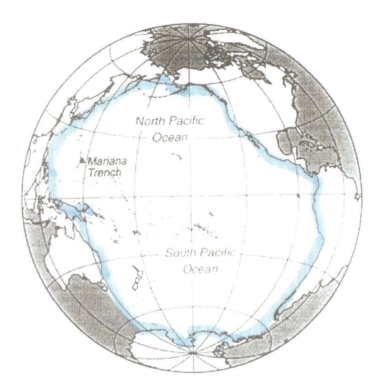

The Pacific Ocean is about 15 times the size of the United States. This ocean covers 33% of the global surface and is larger than the total land area of our planet. Examining an accurate globe of our planet, it is not difficult to imagine that without the Pacific Ocean, our planet could have long ago been gouged by a collision.

The Marianas Trench, 6.5 miles below the surface of the western Pacific, is the lowest known point on Earth. If Mount Everest were located at this site, there would still be one mile of water covering its peak.

Permian Period
225 million years ago

Triassic Period
200 million years ago

Jurassic Period
135 million years ago

Cretaceous Period
65 million years ago

Present Day

Richard Feynman

In the spring of 1950, as my curiosity peaked regarding what I perceived as advanced scientific knowledge appearing in Genesis, I was able to obtain vital insights from my physics professor Richard Feynman.

Dr. Feynman was my most talented teacher. He could explain Einstein's Theory of Relativity in such a dramatic way that I was able to comprehend it. He was an extremely advanced forecaster of the direction that our technology was headed. I remember one lecture wherein he held up his pack of cigarettes and declared that one day the Library of Congress would be stored in a volume equal to that package. He titled that lecture, "Room at the Bottom". He offered us a $10,000 prize if any one of us could build a working electric motor that was no bigger than 1 cubic centimeter. He was so far ahead of today's technology, that his vision of an advanced storage system predicted that information would be stored using molecular layers.

As social chairman of Fleming House at Caltech, I was responsible for inviting faculty members to dine with "the troops" once a month. This is how the opportunity presented itself, for me to have a one-on-one conversation with Dr. Feynman.

My question to him was, "Dr. Feynman, do you believe in UFO's?" His answer, "Klarfeld, I believe in the Law of Probability. Of the billions of stars in our galaxy and the billions of galaxies in the universe, the Law of Probability says there are 10,000 systems just like ours in every detail. Our star is one of the youngest of the 10,000 systems. If any of the 'others' survived their space age, they could have explored us. Yes, I believe in UFO's."

Today, as I have put together the irrefutable physical evidence of the work of the incomprehensibly intelligent species that occupied our planet thousands of years ago, I hark back to Richard Feynman's answer to my UFO question.

"Dr. Feynman, do you believe in UFO's?"

"Klarfeld, I believe in the Law of Probability. Of the billions of stars in our galaxy and the billions of galaxies in the universe, the Law of Probability says there are 10,000 systems just like ours in every detail. Our star is one of the youngest of the 10,000 systems. If any of the others survived their space age, they could have explored us. Yes, I believe in UFO's."

Connecting the Dots

Having convinced myself that the story of the collision between Tiamat and Nibiru's moon "Northwind", as described in the Enuma Elish story, could have resulted in an Earth that lost half of its mass, I set out to collect as much physical evidence that I could find that would prove the presence of the Anunnaki on Earth thousands of years ago.

1. **(Stonehenge)** was my first project. Research has revealed that before the Sarsen Circle of upright stones was erected, a 285-foot diameter circle of 56 chalk holes, 3 feet in diameter, was created. (These are called the Aubrey Holes, in honor of John Aubrey.) Pg.21.

A CBS TV program in the 1960's ran a computer analysis of the Aubrey circle. They declared that Stonehenge's location—latitude 51 degrees 11 minutes, was a very special location for eclipses of the moon. This location produces moon eclipses in the repeating sequence of 19 years, 19 years, and 18 years. Adding 19+19+18=56. Thus if the white 3 foot diameter chalk holes were covered by a black stone, that was moved around the circle in sync with the passage of moon cycles, the black stone would arrive at the heel stone position, on the exact day when a moon eclipse would occur. (Eclipse computer.) (S.I.D.)

How could this stone computer have been created without the precise knowledge of the celestial mechanics of this unique geographic location? Certainly this was not the work of the early tribes that lived on this Salisbury Plain, thousands of years ago.

An Aerial View of Stonehenge

The view includes the circular bank, ditch, and counterscarp bank.
A number of the Aubrey holes are also visible. The Heel Stone can be
seen in the lower right.

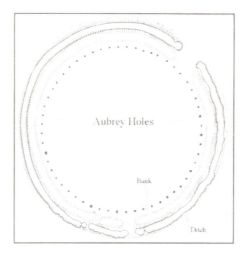

Stonehenge: Phase 1 (2950-2900 BCE)

The earliest portion of the complex dates to approximately 2950-2900 BCE. (Middle
Neolithic). It is comprised of a circular bank, ditch, and the counterscarp bank of about
330 feet (100 meters) in diameter. Just inside the earth bank is a circle of the 56 Aubrey
holes.

This <u>S.I.D.</u> (<u>S</u>tored <u>I</u>nformation <u>D</u>evice) clearly displays enormous information about planet Earth's celestial relationships with the Sun, the Moon and the rotational speed of our planet.

Following the stone computer, came the erection of the 30 upright stones that formed the Sarsen Circle, 100 feet in diameter. My question was why 30? I divided 360 degrees by 30 and discovered the number 12. The number 12 is one of the most important numbers in the Anunnaki civilization...their Pantheon consisted of the Twelve Great Anunnaki gods, they declared 12 months in one year, 2 twelve hour parts of each day, they created the 12 signs of the Zodiac. These Sarsen uprights are harder than granite and weigh 25 tons each. They were quarried at Marlborough Downs using tools not locally available at that time, and then these huge stones were transported over 20 miles to this site.

The final construction placed each upright stone into the ground, forming the 100-foot circle with the top level of all <u>30</u> <u>stones</u> extraordinarily attaining the identical elevation.

The tops of these uprights were linked by a continuous ring of 30 horizontal Sarsen lintels, (Pg. 23), curved to follow the circumference of the circle. These lintels were anchored in place by tenon joints (pin and hole in the vertical plane), and tongue and groove in the horizontal plane. I believe that this is the first manifestation of a temple structure, using tall vertical columns and with caps stones to tie the columns together.

When I realized all of the intricate details that were accomplished in this part of the Stonehenge complex, plus the 45-ton Sarsen Trilithon sighting structures, erected inside of the Sarsen circle, with much heavier lintels, I could only conclude that an enormously advanced intelligence built these structures. <u>S.I.St.</u> (<u>S</u>tored <u>I</u>nformation <u>S</u>tructure).

This construction technique appeared later at many temples in Egypt, Greece and Rome. (Those structures were rectangular, rather than the circular shape of Stonehenge.)

Stonehenge at the end of Phase III

Heel Stone

2. **(Landing Platform)** The colossal Roman ruin at Baalbek, Lebanon was built on a "tel" mound, indicating a place that had long been held sacred. Why the Romans built this largest and tallest temple, to their god Jupiter, 1,500 miles from Rome, is the key question. The Roman construction was built on top of an existing 5,000,000 square foot platform. The scale of the Roman project, so distant from the center of their empire, indicates that this site was extraordinarily sacred.

Zecharia Sitchin's research led him to conclude that the original platform was the major landing site of the Anunnaki. Pg.25 shows an aerial view of the Roman ruins on top of "the landing platform." The six columns with their capstones still attached, are the remains of the Jupiter Temple. The other temple was dedicated to Venus.

The northwest corner of this platform is anchored by <u>3 of the largest stones ever used in any construction on our planet</u>! These stones measure 64 feet long by 14 feet by 14 feet and each weigh over 1,200 tons. (The St. Louis Arch weighs 900 tons.) The builders of the landing platform quarried these megalithic stones, transported them one-half mile up hill to the platform site, lifted them 36 feet on top of the carefully prepared foundation course, and fitted them perfectly end to end. Our best 21st century technology can not duplicate this feat!

The partial quote displayed on Pg. 25, is from Sitchin's recent book, a translation called "The Lost Book of Enki."

If this platform was built pre-flood (olden times), then the design should be classified a survival success. Another super large stone has been found under the Western Wall in Jerusalem. It is possible that Solomon's Temple was built on top of another Anunnaki structure.

This very ancient 5,000,000 square foot platform is an engineering marvel made of stone and assembled without the need of mortar.

The amazing construction achievements presented at Baalbek qualify this platform as a major <u>S.I.St.</u> (Stored Information Structure).

In Jerusalem – The WESTERN WALL MASTER COURSE contains, four large stones. The largest of which weighs 570 tons, and is 44 ft. long, 10 ft. high, and 12-16 ft. deep.

"All that in the Olden Times in the Edin and the Abzu had existed
        under the mud was buried!
Eridu, Nibru-ki, Shurubak, Sippar, all were gone, completely
        vanished;
But in the Cedar Mountains the great stone platform in the sunlight
        glistened,
The Landing Place, in the Olden Times established, was still
        standing!
One after another the Whirlwinds upon the platform landed;

The Platform was intact; at the launch corner the huge stone blocks
        Held firm."

"Where a Landing Place to establish, a place for the rocketships, he
          was seeking.
Enlil, by the heat of the sun afflicted, for a place of coolness and
          shade was searching.
To snow-covered mountains on the Edin's north side he took a liking.
The tallest trees he ever saw grew there in a cedar forest.

There above a mountain valley with power beams the surface he
          flattened.
Great stones from the hillside the heroes quarried and to size cut.
To uphold the platform with skyships they carried and emplaced
          them.
With satisfaction did Enlil the handiwork consider,
A work beyond belief indeed it was, a structure of everlasting!"

Gateway Arch weighs 900 Tons

This stone weighs over 1200 tons

3 megalith stones, raised 36 feet above ground, placed end to end

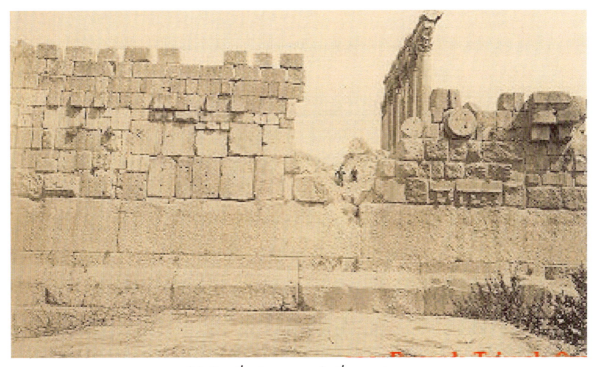

Notice the two men in the center...

3. **(Giza Pyramids)** Shuttle craft re-entering from earth orbit must be carefully programmed to land safely at a predetermined location. Today, NASA has the necessary ground support systems after years of building a huge infrastructure. When our planet was void of any such systems, those who built the landing platform at Baalbek, required navigational landmark beacons necessary to repeatedly locate this landing site.

Pg.29. quotes again from Sitchin's "The Lost Book of Enki," how the two artificial mountains at Giza were created to provide brilliant reflected light during the daytime and artificial night lighting. Using the natural twin peaks of Mount Ararat, a navigational line could be projected between the two Giza Pyramids and Mount Ararat, thus locating the Baalbek platform.

The landing platform at Baalbek was located by the line drawn between Mt. Ararat and the brilliant glow of the Great Pyramid.

© Z. Sitchin
Reprinted with permission

To achieve structural stability for the enormous tonnage of the Pyramids at Giza, a proper foundation was required. A granite mountain top was leveled with a "power beam", similar to the one used to prepare for the Landing Platform (perhaps the Bull of Heaven.) Pg. 30.

Pg. 31 is a painting of early Ciaro showing, on the right hand edge, some very curious small mountains whose tops have been neatly removed. (Over spray?)

One of the most remarkable features of these two Pyramids is that when originally finished, all 8 sides were covered with optically polished casing stones. These mirrored surfaces reflected sunlight with a brilliance that could be seen hundreds of miles in all directions. Pg. 32.

The precision with which the casing stones were made and installed (Pg.33) could only be the work of builders with incomprehensible intelligence. <u>S.I.St.</u> (<u>S</u>tored <u>I</u>nformation <u>St</u>ructure).

THE BEACON

"Where the second set of twin peaks was required, mountain there were none.
Artificial peaks thereon we can raise! So did Ningishzidda to the leader say,
On a tablet the image of smooth-sided, skyward rising peaks for them he drew.
If it can be done, let it so be! Enlil with approval said. Let them also as beacons
        Serve!.......
By the Anunnaki, with their tools of power, were its stones cut and erected.
Beside it, in a precise location, the peak that was its twin he placed;
With galleries and chambers for pulsating crystals he designed it.
When this artful peak to the heavens rose, to place upon it the capstone the leaders
        were invited.

Of electrum, an admixture by Gibil fashioned, was the Apex Stone made.
The sunlight to the horizon is reflected, by night like a pliiar of fire it was,
The power of all crystals to the heavens in a beam it focused."

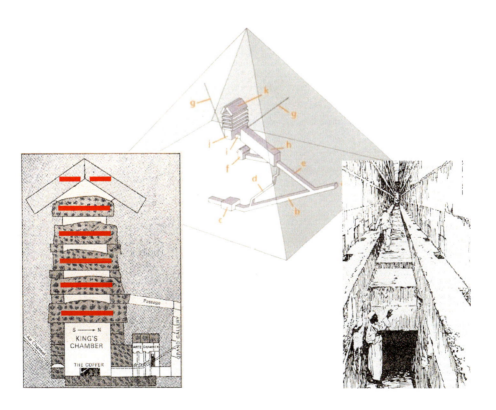

"GUG" pulsating stone in coffer
below 5 hollow compartments...
Page 47 paragraph 5

27 pairs of slotted holes in Grand
Gallery, held radiating crystals
Page 47 paragraph 5

The foundation (a granite mountain top) was leveled to an amazing degree. Each base covers more than thirteen acres, with no corner higher or lower than one-half inch. This near-perfect leveling, far exceeds the finest architectural standards of the present day.

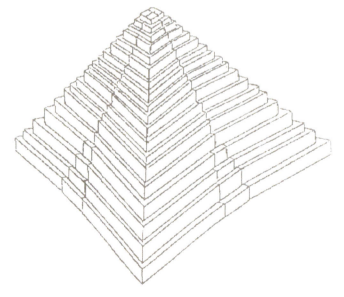

The four faces of the pyramid are slightly concave. The centers of the four sides are indented with extra-ordinary degree of precision. This effect is only visible from the air. The curvature designed into these faces of the pyramid exactly matches the radius of the earth.

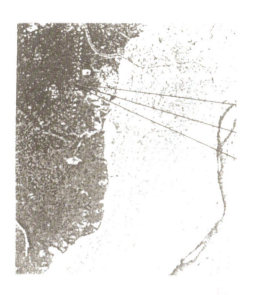

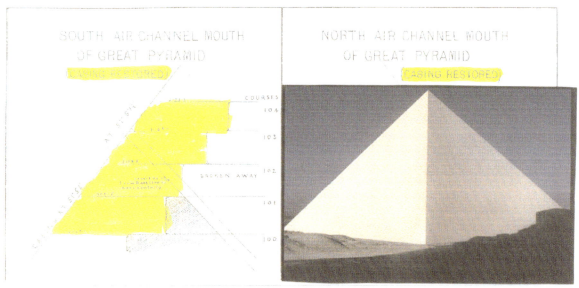

SOUTH AIR CHANNEL MOUTH OF GREAT PYRAMID (CASING RESTORED)

NORTH AIR CHANNEL MOUTH OF GREAT PYRAMID (CASING RESTORED)

The Great Pyramid was originally encased with smooth white optically polished limestone blocks. These stones were polished on all six sides, to a tolerance of 0.01 inch. The resulting 13-acre mirror surfaces reflected sunlight to every horizon, creating the most overwhelming sight on Earth. Legendary Egyptologist Sir William Flinders Petrie expressed his astonishment at this feat by writing, "Merely to place such stones in exact contact would be careful work, but, <u>to do so with cement in the joint seems almost impossible</u>; it is to be compared to the finest opticians' work on a scale of acres." Hundreds of these casing stones are still glued at the top of the middle pyramid, <u>proof of the Incredible adhesive power of this glue</u>, and its durability over the past 10,000 years! Although it has been tried, <u>no one has been able to replicate this glue today</u>!

THE GEOPOLYMER LIBRARY
Publications Abstracts (part 2)

X-Ray Analysis and X-Ray Diffraction of Casing Stones from the Pyramids of Egypt, and the Limestone of the Associated Quarries (1984), by Joseph Davidovits

Published in Science in Egyptology, Proceedings of the Science in Egyptology Symposia, Manchester, U.K., pp. 511-520, 1984.

The hypothesis that the limestone that constitutes the major pyramids of the Old Kingdom of Egypt is man-made stone, is discussed. Samples from six different sites at the traditionally associated quarries of Turah and Mokattam have been studied using thin-section, chemical X-Ray diffraction. The results were compared with pyramid casing stones of Cheops, Teti and Seneferu. The quarry samples are pure limestone consisting of 96-99% Calcite, 0.5-2.5% Quartz, and very small amount of dolomite, gypsum and iron-alumino-silicate. On the other hand the Cheops and Teti casing stones are limestone consisting of: calcite 85-90% and a high amount of special minerals such as Opal CT, hydroxyl-apatite, a silico-aluminate, which are not found in the quarries. The pyramid casing stones are light in density and contain numerous trapped air bubbles, unlike the quarry samples which are uniformly dense. If the casing stones were natural limestone, quarries different from those traditionally associated with the pyramid sites must be found, but where? X-Ray diffraction of a red casing stone coating is the first proof to demonstrate the fact that a complicated man-made geopolymeric system was produced in Egypt 4,700 years ago.

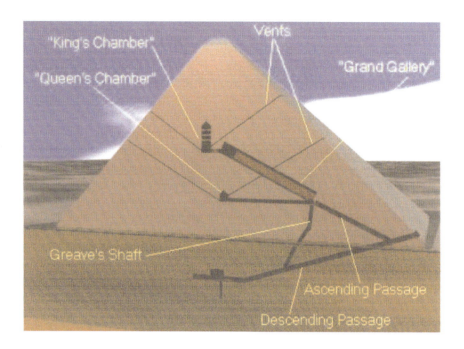

The Great Pyramid is the most substantial ancient structure in the world. Originally 481 feet, 5 inches tall and measuring 755 feet along its sides, the base covers an area of 13 acres. Constructed from approximately 2.5 million limestone blocks weighing on average 2.6 tons each, its total mass is more than 6.3 million tons. This is more building material than is found in all the churches & cathedrals built in England since 1 A.D.

Additional facts indicating that the builders were <u>NOT</u> the Egyptians:

1. The massive granite stones in the inner chambers, have a Mohs scale mineral hardness of 5 to 6. The copper and bronze tools used by the Egyptians had a hardness of 3.5 to 4. The rectangular passageways penetrating deep into the <u>granite</u> foundation, and the interior granite coffer, could not have been produced with copper or bronze tools.

2. The foundation (a granite mountain top) was leveled to an amazing degree. Each base covers more than thirteen acres, with no corner higher or lower than one-half inch. This near-perfect leveling, far exceeds the finest architectural standards of the present day.

3. Measurements throughout the pyramid show that its constructors knew of the proportions of pi (3.14...), and phi or the Golden Mean (1.618), and "Pythagorean" triangles, thousands of years before Pythagoras lived.

4. Measurements of each base show that the builders knew the precise spherical shape and size of Earth, Venus, Mars, and Mercury.

Today, 10,000 plus years later, hundreds of these remarkable casing stones are still glued at the top of the second pyramid, demonstrating the incredible adhesive power of the glue. As Egyptologist Sir William Flinders Petrie stated, "Merely to place such stones in exact contact would be careful work, but to do so with cement in the joint seems almost impossible; it is to be compared to the finest opticians' work on the scale of acres".

The possibility that the builders manufactured the casing stones (Pg. 33) as the analysis by Joseph Davidovits states, only adds to the body of evidence that the builders of the pyramids possessed extraordinary skills. Skills exceeding those possessed by current 21st century technology.

Pg. 34 is a list of a few more of the outstanding facts about the dimensions and construction features that are incorporated in these monuments of knowledge. Not content to provide themselves with navigational beacons, the builders stored enormous volumes of scientific and mathematical information for future civilizations to decipher. Giza Pyramids are S.I.St. (Stored Information Structure)

4. **(Easter Island)** Those orbiting planet Earth thousands of years ago required additional visual navigational aides to work in conjunction with the Beacons at Giza. Pg. 36; Ilustrates the "night time-day time" cycle earth orbiters encounter. An ideal system should include available reflective beacons on every revolution. Easter Island is located almost 12 hours ahead of Giza. The statues that ring the perimeter of Easter Island, all face east, with their faces slightly up turned. The polished white coral eyes that were found on these statues, should have reflected the suns rays vertically to produce a second "Beacon." The top hats could also reflect the sun if polished white coral had been applied to these surfaces. Some of these statues were anchored atop of cyclopean walls. This type of construction was favored by the builders of the Baalbek platform and the Giza Pyramids. The discovery of this type of masonry on Easter Island is evidence of whom the statue builders were. (Human ears lengthen during the aging process. All Easter Island statues have extra long ears, big noses, and beards….all features of the Anunnaki.)

Using statues to produce night time light is illustrated on Pg.37. The Atlantes statues appear to depict a row of robots that were armed with gas-fired devices at the extremes on their right side. These jet-gun like tools were depicted to produce flames from men with back packs. (See center sculpture.) If the "real life" robot warriors were stationed in a line, and signaled to produce fire (light at night landings), the right arms could have pivoted out and upward until their arms were locked in an upright position (held in place by the huge magnets that surrounded their right ears.) Their chest plates and backside shields could have held the fuel. (S.I.D. and S.I.Sc.)

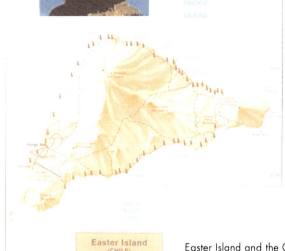

Plain of Jars

Located in Laos, on the eastern edge of the Asian continent, outlined by the Mekong River, over 1000 2.5 ton granite stone jars have been mysteriously pointing skyward for 1000's of years. Could these containers, (made from material not found in Laos), be an additional "Beacon" used by landing shutlecraft?

Easter Island and the Giza Pyramids were Earth's first visible navigational Beacons. Landing sites were at Baalbek and Nazca.

At the northwest corner of France, the town of Carnac contains 3000 Megalith Stones in parallel rows, that may have served as an additional visual navigation array for orbiting space ships.

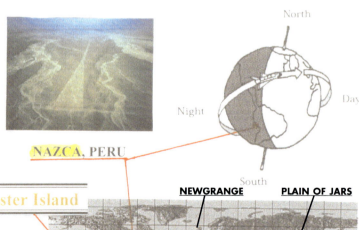

NAZCA, PERU

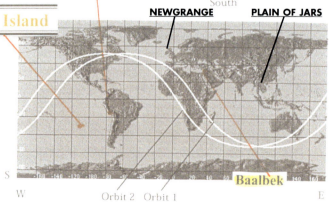

NEWGRANGE, IRELAND

**This 5200 year old construction has all of the requirements of an
Anunnaki earth navigational "Beacon". The astronomical feature,
(the Sun shines down the central passage during the Winter Solstice)
plus the 1000's of shimmering white quartz stones that cover the side
and part of the roof of this structure, make it an ideal shiny landmark
that can be seen from an orbiting spaceship. (White quartz stones and
the round dark granite stones are not found locally at this site).**

Winter solstice sun illuminates the mound at Newgrange.

5. **(Cyclopean Walls)** The oldest discovered use of stone structures, known as cyclopean walls, was found at Baalbek, Lebanon. This massive platform (5,000,000 sq.ft.) was built using four layers of perfectly cut stones (no masonry). The top level in the northwest corner consisted of the three largest stones ever used in any construction project on the planet. These megalithic stones (each weighing 1,200 tons) were precisely placed on top of the foundation's courses. The survivability of this type of stone arrangement is a colossal testimony to the skill and knowledge the builders. Sitchin's translations indicate that the landing platform was built pre-flood (over 10,000 years ago.) Pg. 39.

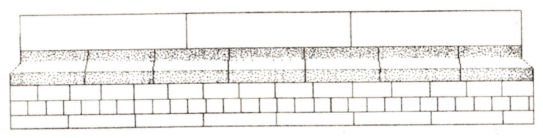

© Gian J. Quasar

Using this same technique (stone structures that fit together with extraordinary precision) walls were created at some later date at Tiahuanaca, Machu Picchu, Cusco, Sacuahuaman and Ollantaytambu. Pg. 41 and Pg. 42 illustrate the extraordinary workmanship that produced these beautiful walls, with perfect seams, without mortar. Many of these stone structures were celestial observatories.

The stones enclosed in the red circle on Pg. 42 have raised my curiosity about the possibility that these stones were manufactured. Natural stones usually do not contain impressions of unnatural shapes. Note the double protrusions at the bottom of many of the "stones" below at Ollantaytambo. Could these be the remaining signs of an injection process used to create these "stones?"

Ollantaytambo
Photo by Kurt Bennett

6. **(New World Pyramids)** The exile of Thoth from Egypt in 3113 B.C. coincides with the arrival in Mesoamerica of Quetzalcoatl. Pg. 44. The start date of the Calendar Round is 3113 B.C. This slide rule type calculator was capable of producing extremely accurate dates over thousands of years. The tracking of time by calendar was and is a mark of advanced civilization. The end date for this period is December 22, 2012. (<u>S.I.D.</u> <u>S</u>tored <u>I</u>nformation <u>D</u>evice).

Z. Sitchin credits the Egyptian god Thoth (son of ENKI) as the designer of the Giza Pyramids. Quetzalcoatl is described as quite different physical type from the natives--fair skinned and ruddy complexion, long nosed and with a long beard. Quetzalcoatl is credited with bringing civilization, learning, culture, the calendar, mathematics, metallurgy, astronomy, masonry, productive agriculture, knowledge of healing, properties of plants, law, crafts, and the arts to the New World. (This list of firsts for Mesoamerica reads like the "gifts" given the Sumerians. Pg. iii). The one missing "gift", the wheel, given to the Sumerians, appears to have been withheld. (I speculate that without the wheel, human population was easier to control.)

In addition to bringing civilization, Quetzalcoatl built pyramids. The three major pyramids at Teotihuacan resembles the Giza Orion Belt array. Pg. 44 The base dimension of the Pyramid of the Sun is close to that of the Great Pyramid at Giza. The major modification in style of the New World pyramids is the step construction and stairways on the outside, giving access to the temples at the top of each structure. At Chichen Itza the entire pyramid acted as a solar observatory. The summer solstice was celebrated when the rays of the sun played on the edge of this pyramid, producing a slithering snake like appearance.

The first recorded civilization in Mesoamerica was named after The Olmec people. Enormous (10 feet tall, 25 tons) African head sculptures have been found. Thoth (Quetzalcoatl) was exiled from Egypt, he brought with him very tall African helpers. It is difficult to find other explanations for these ancient African heads.

Quetzalcoatl, while symbolized as a feathered serpent, appears also to have been an historic figure-- the man credited with bringing civilization, learning, culture, the calendar, mathematics, metallurgy, astronomy, masonry, architecture, productive agriculture, knowledge of the healing properties of plants, law, crafts, the arts and peace to the native people. He is pictured as a quite different physical type than the natives--fair skinned and ruddy complexioned, long nosed, and with a long beard. He was said to have arrived by boat from the east, and sailed off again years later promising to return someday.

Quetzalcoatl= THOTH, Son Of ENKI

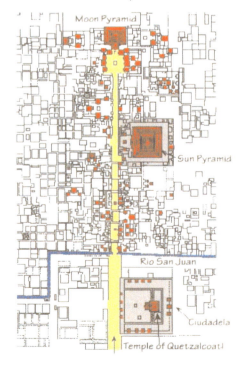

PYRAMIDS OF MEXICO

PYRAMID OF THE SUN

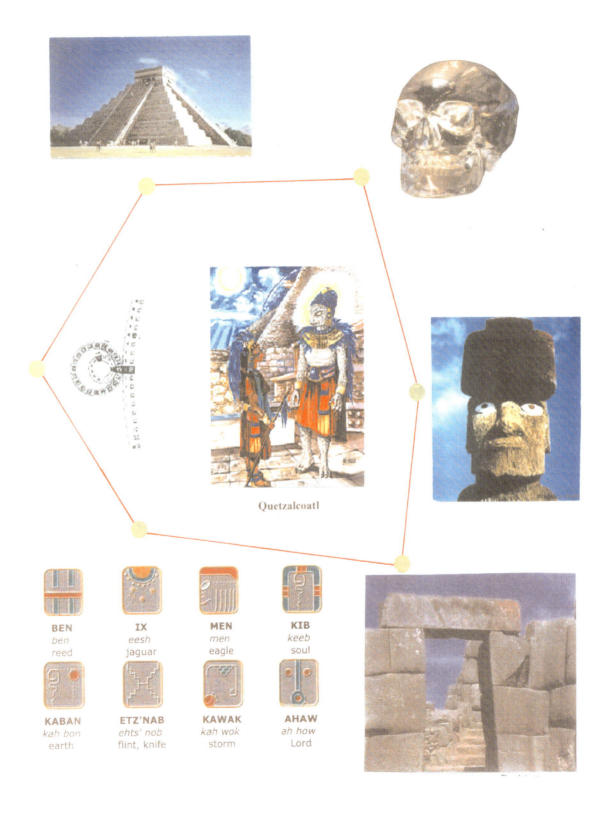

Quetzalcoatl

BEN
ben
reed

IX
eesh
jaguar

MEN
men
eagle

KIB
keeb
soul

KABAN
kah bon
earth

ETZ'NAB
ehts' nob
flint, knife

KAWAK
kah wok
storm

AHAW
ah how
Lord

7. **(Crystal Skull)** In 1927, one of the most famous archeological discoveries from Central America was reported to have been found by Frederick Mitchell-Hedges and his daughter Anna. As shown on page 48, this life-size skull and its detachable jaw are both made from the same solid block of crystal. According to expert mineralogists, rock crystal can only be carved with diamond-tipped tools.

Extensive tests conducted at Hewlett-Packard laboratories in 1970, confirmed that the Crystal Skull was indeed formed from a solid block of quartz crystal. One test conclusively proved that it was made of the purest type of quartz found on Earth. The detachable jawbone was carved from exactly the same piece of rock. Because of the quartz crystal's hardness, the H-P team declared that this skull and its jaw could not have been carved using any known type of modern diamond-tipped power tool. The vibration, heat, and friction generated undoubtedly would have shattered the delicate lower jaw. Normal carving is accomplished by working along these large stones' natural axis, to preserve their structural integrity and thus maximize optical purity with minimum risk of damage from vibration and heat stress. Amazingly, this skull was produced with complete disregard of this natural axis.

Under extreme magnification of the Crystal Skull's surface, no evidence of tool markings or surface scratch marks was detected. Following these examinations one member of the H-P team commented, "This skull shouldn't even exist."

The H-P team further found that the skull's single piece of natural quartz was 'piezo-electric' silicon dioxide, the same type of quartz used in all modern electronics. Piezo-electric silicon dioxide has natural positive and negative polarity, so in a completely literal sense the Crystal Skull is like a battery. Squeezing the skull will generate a small electrical charge.

Z. Sitchin provides a glimpse into the little-understood ancient uses of crystals as revealed in his translations regarding the Great Pyramid at Giza. In "The Pyramid Wars" (page 169), "..many-hued glows were emitted by twenty-seven pairs of diverse crystal stones that were evenly spaced along each side of the (Grand) Gallery…each crystal stone emitted a different radiance…in the uppermost Grand Chamber was the "GUG" pulsating stone…its emissions amplified by five hollow compartments constructed above the chamber…" see Pg.29

We do not know to what end the crystal's relative spacing, radiance, and amplification were directed. But it defies rational thinking to imagine that this precisely configured channeling of light energy was a mere dalliance by the Anunnaki in the grand design of the Great Pyramid!

In "The Mystery of the Crystal Skulls" published by Bear & Co., Rochester, Vermont, Chris Morton and Ceri Louise Thomas present many intriguing future possibilities surrounding the Crystal Skull. One is the Mayan legend that one day The Skull will speak and reveal volumes of stored information necessary to survival of the human species on Earth.

As we have seen in the case of cylinder seal VA/243, refined (S.I.D) forms are complemented by marvelous content we have not yet fully understood. Their refined aesthetics, ingenious forms, and effective storytelling power - recognized in our modern world as requiring higher order intelligence and skill - signify a serious intention to communicate vital information. If the Crystal Skull likewise contains information of impact worthy of its awesome "vehicle", the moment of its revelation promises to be a landmark in human (even cosmic) history.

MITCHELL-HEDGES CRYSTAL SKULL

Skull hinges on jaw sockets.
When suspended from above (two tiny holes are drilled on each side of the base at perfect balance points) the slightest breeze causes the skull to nod back and forth, the jaw opening and closing as counter-weight. The visual effect is that of a living skull, talking.

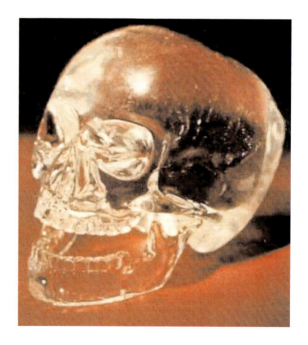

8. **(Nazca Lines)** On the Peruvian desert, 200 miles south of Lima, there is a plateau that is approximately 37 miles long and 1 mile wide. In the 1920s airline pilots flying over this mountainous plain discovered an amazing array of 70 animal and plant figures that include a spider, hummingbird, monkey and a head with two hands. The lines were created by clearing the darkened pampa stones to either side, and exposing the lighter sand underneath. All figures were created by a singular continuous line, which implies careful design and even more amazing execution. The spider is identified as the rare genus Ricinulei found only in the remote Amazon Jungle. Both the Head and the Monkey have 9 fingers. In addition to the 70 Biomorphs there are about 900 Geoglyphs. Geoglyphs are geometric forms that include straight lines, spirals, triangles, circles and trapezoids. They are enormous in size with the longest straight line extending over nine miles across the plain.

One group of straight lines, bursting from a central point, has the same appearance as the celestial map that Carl Sagan put on NASA's Pioneer 10 Plaque and the Voyager 1&2 Golden Records (Page 9). As noted by NASA, using the position of our Sun relative to 14 Pulsars and the center of our Galaxy, the interceptors of our message will know from whence it came. Does this celestial map, drawn on the plateau at Nazca, indicate the origin of its creators? Do the three line intersection clusters echo the Orion Belt formation? S.I.Sc. (Stored Information Sculpture).

From Pg.e.

POSITION OF SUN
RELATIVE TO 14
PULSARS AND THE
CENTER OF THE GALAXY

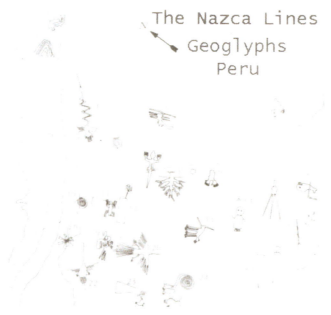

Geoglyph Map of the Nazca Pampa

1. Killer whale 2. Wing 3. Baby Condor 4. Heron Bird 5. Animal 6. Spiral 7. LIzard 8. Tree 9. Hands 10. Spiral 11. Spider 12. Flower 13. Dog 14. Astronaut 15. Triangle 16. Whale 17. Trapezoids 18. Star 19. Pelican 20. Condor Bird 21. Trapezoid 22. Hummingbird 23. Trapezoid 24. Monkey 25. Llama 26. Trapezoids

9. **(Winged Pilot)** Ancient wall sculptures have been found that depict a bearded pilot, with a prominent nose, appearing to be in control of a flying machine. He grasps a circular ring in his left hand and extends his right hand in an open gesture. The large circular ring that encases his lower body can be interpreted as the "sphere" that was used in ancient drawings of The 10th Planet. The two lower protruding struts could be the landing gear for this flying machine. The extremely straight leading edge of this sculpture can not be a replication of a bird's wing, as nature does not produce birds with totally straight leading edge wing spans.

Records of Sumerian civilization referred extensively to "The Lords who came down from above". They called the home planet of these "Lords", NIBIRU. They also named NIBIRU, the planet of the crossing. Other civilizations also experienced "The 10th Planet" and attached other names; Babylon-MARDUK, Hebrew-WINGED GLOBE, Hindu-TRETA YUGA, Egyptian-CELESTIAL DISK, and Greek-NEMESIS. Because of the 3,600-year orbital cycle, the Romans never experienced "The 10th Planet." Winged Pilot. S.I.Sc.

10. **Life size statue of Ishtar**. (Pg. 5) S.I.Sc.

11. **Ancient Egyptian Flying Vehicles**. (Pg. 7) S.I.Sc.

12. **Cylinder Seal VA/243**. (Pg. 8) S.I.D.

These 12 Stored Information objects are real and can be touched, studied, and recognized as creations of an incomprehensible advanced technology. When all 12 of these examples are examined as a group, it absolutely convinces me that they represent IRREFUTABLE EVIDENCE that the Anunnaki did occupy our planet thousands of years ago. The translations in the Sumerian cuneiform tablets are not fables. The pictures (cylinder seal CA/243) are self portraits of Anunnaki leaders. The life-size statue of Ishtar is an actual reproduction of this beautiful Goddess.

Confirming the Anunnaki's occupation, my next study is the examination of the Theory of Evolution........ Are Homo sapiens a genetically engineered species?

THE 10TH PLANET

SUMER   NIBIRU   (Planet of the Crossing)

BABYLYON   MARDUK   (King of the Heavens)

MESOPOTAMIAN   MARDUK   (The Great Heavenly Body)

HEBREW   WINGED GLOBE

HINDU   TRETA YUGA

EGYPTIAN   CELESTIAL DISK

GREEK   NEMESIS

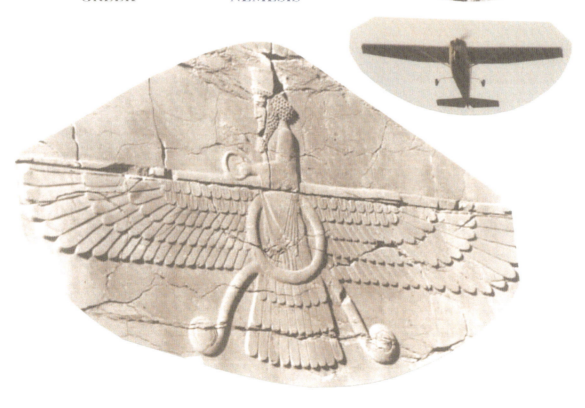

ADAM

The Book of Genesis bears a very strong echo to Enuma Elish:

Enuma Elish: "When on high the heaven had not been named, firm ground below had not been called by name...."

Genesis: "In the beginning God created the heaven and the earth. And the earth was without form, and void; and darkness was upon the face of the deep."

As I stated in Chapter 3, the story of Enuma Elish was written on seven tablets. The creation process took six tablets to complete, and the seventh tablet deals with "the Creator" exalting in his handiwork and the greatness of his creation. In Genesis 2.2 "He rested."

Genesis 1.26 "And God said, let Us make man in Our image, according to Our likeness."
Genesis 2.7 "And God formed the man out of dust from the ground..."
Genesis 2.21 "And God caused a deep sleep to fall on man, and he slept. And He took one of his ribs, and closed up the flesh underneath."
Genesis 2.22 "And God formed the rib which He had taken from the man into a woman, and brought her to the man."

In light of today's cloning technology, Genesis 1.26-27 reads very much as if it was taken from Lord Enki's autobiographical story as translated by Zecharaia Sitchin in "The Lost Book of Enki" tablet six, page 139-140.

"Enki the boy child held in his hands; The image of perfection he was....
Now this is the account of how Adam by name was called,
And how Ti-Amat as a counterpart female for him was fashioned.
The newborn's visage and limbs the leaders carefully examined:
Of good shape were his ears, his eyes were not clogged,
His limbs were proper, hindparts like legs, foreparts like hands were shaped.

Shaggy like the wild ones he was not, dark black his head hair was,
Smooth was his skin, smooth as the Anunnaki skin it was,
Like dark red blood was its color, like the clay of the Abzu was its hue,
They looked at his malehood: odd was its shape, by a skin was its forepart surrounded,
Unlike that of Anunnaki malehood it was, a skin from its forepart was hanging!

Let the Earthling from us Anunnaki by this foreskin be distinguished! So was Enki saying.
…A name will you give him? Enki inquired. A Being he is, not a creature!
Ninmah cast her hand upon the newborn's body, with fingers his dark red skin she caressed.
Adam I shall call him! Ninmah was saying. One Who Like Earth's Clay Is, that will be his name!
For the newborn Adam a crib they fashioned, in a corner of the House of Life they placed him.
A model for Primitive Workers we have indeed attained! Enki was saying.
Now a host of Workers like him are needed!…"

This description of the final success, after many failures, to create a "primitive worker" by cloning Anunnaki DNA onto the local population of hairy primitive ape like creatures (Homo erectus), is very strong evidence of the jump start of our species, Homo sapiens. Homo erectus was not suitable as a "primitive worker" probably because it was unable to speak. The work was done in east central Africa, adjacent to the Anunnaki gold mining operation. (Archeology research point to this location as the origins of our species.)

The human genome study revealed that humans (Homo sapiens) possess approximately 25,000 genes. They also discovered that, of the human genome total genes, 223 genes do not have the required predecessors on the genomic evolutionary tree. The question now confronting this study's is, "How did these alien genes become a part of human DNA?" My answer is to accept the fact that we are a genetically engineered species. The S.I.T. (Stored Information Text) tells us exactly how it was done, where it was done, and when it was done. (The echo in Genesis 1.26-2.22 now must confront this revelation.)

Neil Freer, author of "Breaking the Godspell", published a White Paper, October 6, 1999, wherein he addressed "The Alien Question."

…"The documentary evidence, i.e. the historical documentation for the existence and deeds of the Anunnaki, has been available to us since the early 1800's. The excavation of the ancient sites of Mesopotamia brought to light the amazingly advanced civilization of Sumer and, with it, thousands of clay tablets containing not only mundane records of commerce, marriages, military actions and astronomical calculation systems but of the history of the Anunnaki themselves. It is clear from these records that the Sumerians knew these aliens to be real flesh and blood. The library of the ruler, Ashurbanipal at Nineveh was discovered to have burnt down and the clay tablets held there were fired, preserving them for our reading. Even to this day, more and more records are discovered.

One of the most impressive finds, in very recent time, has been a sealed, nine-by-six foot room in Sippar holding, neatly arranged on shelves, a set of some 400 elaborate clay tablets containing an unbroken record of the history of those ancient times, a sort of time capsule. Again, the evidence is so overwhelming and robust that, if it weren't for those with power enough to suppress, it would have been accepted and our world view changed a century ago or perhaps sooner.

The recovered records place the location of the Anunnaki laboratory where the first humans were literally produced in east central Africa just above their gold mines. This falls precisely on the map where the mitochondrial DNA "search for Eve" places the first woman homosapiens and in the same time frame. The evidence of advanced genetic engineering is all there in the ancient documents. Our rapid progress from inception to going to Mars soon, after only 250,000 years, does not correspond to the million-year periodicities of slow evolutionary development of other species such as homo-erectus before us. The HH paradigm shows that the Creationists were only half wrong and the Evolutionists only half right: there was a creation event but it was a genetic engineering process; there is an evolutionary process but it was interrupted in our regard by the Anunnaki for their own practical purposes.

Working from the same archaeological discoveries, artifacts, and recovered records as archaeologists and linguists have for two hundred years, Z. Sitchin propounds – proves, in the opinion of this author – that the Anunnaki (Sumerian: "those who came down from the heavens"), an advanced civilization from the tenth planet in our solar system, splashed down in the Persian gulf area around 432,000 years ago, colonized the planet, with the purpose of obtaining large quantities of gold. Some 250,000 years ago the documents tell us their lower echelon miners rebelled against the conditions in the mines and the Anunnaki directorate decided to create a creature to take their place. Enki, their chief scientist, and Ninhursag, their chief medical officer, after getting no satisfactory results splicing animal and homo-erectus genes, merged their Anunnaki genes with that of Homo-Erectus and produced us, Homo sapiens, a genetically bicameral species, for their purposes as slaves. Because we were a hybrid, we could not procreate. The demand for us as workers became greater and we were genetically manipulated to reproduce.

Eventually, we became so numerous that some of us were expelled from the Anunnaki city centers, gradually spreading over the planet. Having become a stable genetic stock and developing more precociously than, perhaps, the Anunnaki had anticipated, the Anunnaki began to be attracted to humans as sexual partners and children were born of these unions.

This was unacceptable to the majority of the Anunnaki high council and it was decided to wipe out the human population through a flood that was predictable when Nibiru, the tenth planet in our solar system and the Anunnaki home planet, came through the inner solar system again (around 12,500 years ago) on one of its periodic 3,600 year returns. Some humans were saved by the action of the Anunnaki, Enki, who was sympathetic to the humans he had originally genetically created. For thousands of years we were their slaves, their workers, their servants, their soldiers in their political battles among themselves. The Anunnaki used us in the construction of their palaces (we retro-project the religious notion of temple on these now), their cities, their mining and refining complexes and their astronomical installations on all continents. They expanded from Mesopotamia to Egypt to India to South and Central America and the stamp of their presence can be found in the farthest reaches of the Planet.

Around 6,000 years ago, probably realizing that they were going to phase off the planet, began to gradually bring humans to independence. Sumer, a human civilization, amazing in its "sudden" and mature and highly advanced character was set up under their tutelage in Mesopotamia, human kings were inaugurated as go-betweens, foremen of the human populations answering to the Anunnaki. Some humans were taught technology, mathematics, astronomy, advanced crafts and the ways of civilized society. The high civilizations of Egypt and Central America arose.

The Anunnaki became somewhat more remote from humans. By around 1250 B.C. they had gone into their final phase-out mode. The human population and the foremen kings, now left on their own, began to fend for themselves. For some three thousand years, subsequently, we humans have been going through a traumatic transition to independence. Proprietary claims made by various groups of humans as to who knew what we should be doing to get the Anunnaki to return or when they returned, perpetuated the palace and social rituals learned under the Anunnaki, and sometimes disagreement and strife broke out between them. Religion, as we know it, took form, focused on the "god" or "gods", clearly and unambiguously known to the humans who were in contact with them as imperfect, flesh-and-blood humanoids, now absent. It was only much later that the Anunnaki were eventually sublimated into cosmic character and status and, later on, mythologized due to remoteness in time."

This concludes the excerpts from Neil Freer's "The Alien Question: An Expanded Perspective – A White Paper" Posted Wed. 6-Oct.1999 04:38:17 GMT.

On page 58 I have assembled "The Family of Man" tableau. For 6,000 years all major civilizations of the world (Sumer, Babylon, Egypt, Greece and Rome) worshiped basically the same 12 "gods."

Every Anunnaki leader is portrayed on cylinder seals with a "horned" headdress. The life-size statue of Ishtar displays her space helmet neatly adorned with horns. Egyptian images of RA show horned headdress plus fake beards. The horn symbol of leadership was maintained in every major civilization. (Greeks and Romans adopted golden head wreaths.)

It started with the Anunnaki and even appeared in the Bible when Moses came down from the mountain....he had been given the leadership symbol....horns.

ENLIL   NINHARSAG   ENKI

The Family of Man

ISHTAR   NINKI holding ADAM

ANU, Father of the Gods   RA, Egyptian SUN God

APHRODITE/VENUS   MOSES   ZEUS/JUPITER,

LINUS PAULING

In Chapter 4, I recalled my one-on-one meeting with Richard Feynman. In a similar meeting that year, I met with my chemistry professor Linus Pauling.

Dr. Pauling was my most memorable teacher. During my chemistry course, he was using our class to proof his new chemistry text book. One day he started his lecture by lighting a Bunsen burner under a beaker containing clear liquid. He then directed us to page 147 of his book. Very quickly he turned his back to the class and wrote the equation from page 147 on the blackboard. Finishing his famous scribble, he faced the class and told us that he had dreamt about this formula last night. He told us to please put a square root of 2 in front of this formula. In his typical humorous manner he announced that the square root of 2 was the best "fudge factor" that he knew about and since we lived in an imperfect world, he felt it was necessary to modify this formula. With great flourish, Dr. Pauling scratched the square root of 2 on the blackboard, turned toward us, reached into his jacket pocket and with his wonderful mischievous smile, removed a tea bag and dunked it into the now boiling water in the beaker. My perfect black and white world was forever modified!

The night that Dr. Pauling was our guest at dinner, I asked him,
"Dr. Pauling, do you believe in God?"
"Klarfeld, our discipline is to explain everything back to the beginning of the universe. If you ask me what there was just before the Big Bang, I cannot explain that to you. If you wish to believe in a God as the creator, please do, as we can not explain what there was just a millisecond before the Big Bang."

I did choose the option of God as the creator, but I think of Him as the Anunnaki's "Creator of All."

CALTECH APRIL 1950

"Dr. Pauling, do you believe in God?"

"Klarfeld, our discipline is to explain everything back to the beginning of the universe. If you ask me what there was just before the Big Bang, I cannot explain that to you. If you wish to believe in a God as the creator, please do, as we can not explain what there was just a millisecond before the Big Bang."

CONCLUSIONS

As I finish this story I am watching the forward progress of United States tanks and armored vehicles moving across the desert from the Persian Gulf toward Bagdad. When Enki and his group of Anunnaki heroes landed in the Persian Gulf, some 400 thousand years ago, they established their original colony along this very same route. Enki named the first settlement ERIDU ("house in faraway built"). Later, the Sumer civilization, built their cities of Uruk, Nippur, Kish, Larsa, Ur, and Eridu in the valleys of the Tigris and Euphrates rivers.

In the 11th tablet of the Epic of Gilgamesh we learn that Enki's dream message (delivered by Galzu, messenger from The Creator of All) told him that it was the destiny of the Anunnaki to save humanity. He was given detailed instructions on how he was to accomplish this achievement—(described on Page 4). Following this pivotal event, the Anunnaki undertook major efforts to advance human civilization. The results are the historical records of this spectacular first civilization with all of its accomplishments.

In the 14th tablet of Zecharia Sitchin's "The Lost Book of Enki" page 315-318, there is a dialog between Enki and his half brother Enlil in which they debate the use of the Weapons of Terror. Seven atomic tipped missiles had been brought to earth in the earlier times. As disputes grew between the brothers' clans and a power struggle finally erupted, these Weapons were used (Sodom & Gomorrah) and the "evil wind" (radiation fall out) drifted over Sumer totally destroying this first civilization.

"Babli, where Marduk supremacy declared, by the Evil Wind was spared;
All the lands south of Babili the Evil Wind devoured, the heart of the
Second Region it also touched.
When in the aftermath of the Great Calamity Enlil and Enki to survey the havoc met,
Enki to Enlil the sparing of Babili as a divine omen considered.
That Marduk to supremacy has been destined, by the sparing of Babili is
Confirmed! So did Enki to Enlil say.
The will of the Creator of All it must have been! Enlil to Enki said.

"…To his brother's words Enki listened, his head up and down he nodded.
The First Region is desolate, The Second Region is in confusion, The Third Region is wounded, The Place of the Celestial Chariots is no more; that is what has happened! Enki to Enlil said If that was the will of the Creator of All, that is what of our Mission to Earth remained!…

"I what I did did, you what you did did. The past undone cannot become!
Is not in that too a lesson? Enki asked them both.
Is not what on Earth happened, what on Nibiru had taken place mirrored?
Is not in that tale of the Past the outline of the Future written—
Will Mankind, in our image created, our attainments and failures repeat?

"Enlil was silent. As he stood up to leave, Enki to him his arm extended.
Let us lock arms as brothers, as comrades who together challenges on an alien planet confronted!
So did Enki to his brother say.
And Enlil, grasping his brother's arm, hugged him as well.
Shall we meet again, on Earth or on Nibiru? Enki asked.
Was Galzu right that we die if we to Nibiru go? Enlil responded.
Then he turned and departed.
Alone was Enki left; only by the thoughts of his heart was he accompanied.

"How it all began and how it thus far ended, he sat and pondered.
Was it all destined, or was it fate by this and that decision fashioned?
If Heaven and Earth by cycles within cycles regulated,
What had happened will again occur? Is the Past---the Future?
Will the Earthlings the Anunnaki emulate, will Earth relive Nibiru?
Will he, the first to arrive, the last to leave be?

"Besieged by thoughts, Enki a decision made:
All the events and decisions, starting with Nibiru to this day on Earth,
To put in a record, a guide to future generations to become;
Let posterity, at a time by destiny designated,
The record read, the Past remember, the Future as prophecy understand,
These are the words of Enki, Firstborn of Anu of Nibiru."

The Universe is perceivable to humanity because we (Homo sapiens) are the only species on earth capable of seeing and trying to understand it. If the Anunnaki are still members of our solar system, then the Universe is observable to them also. Dr. Feynman said that if a civilization "survives their space age" then they could explore the "stars". My opinion is that our destiny is to replicate humanity in the Universe.

POSTSCRIPT

In the translations of cuneiform tablets there are references to Destiny, NAM and to Fate. NAM.TAR Destiny is "what is going to happen will happen" (unalterable), ie. Our planet will orbit the sun until the sun explodes. Fate is the attempt to alter destiny. To me Destiny is the design of a greater power.

The dialog between the two brothers (pages 61-62) is an excellent example of how two powerful Anunnaki submitted to the will of their deity (The Creator of All) when deciding how and why things happened....the sparing of Babili (Babylon) from the evil wind (caused by their use of atomic weapons) was destiny.

The use of atomic weapons appears to be an experience that happened on their home planet..."Is not what on Earth happened, what on Nibiru had taken place mirrored?
Is not in that tale of the Past the outline of the Future written----
Will Mankind, in our image created, our attainment and failures repeat?...
Was it all destined, or was it fate by this and that decision fashioned?
If Heaven and Earth by cycles within cycles regulated,
What had happened will again occur? Is the Past----the future?
Will the Earthlings the Anunnaki emulate, will Earth relive Nibiru?"

We Earthlings are on the threshold of cloning ourselves. It is not unreasonable to imagine that in the future we will need to find a new planet outside of our solar system. It has been calculated that our sun's end (end of days) will occur in about 5 billion years. At the accelerated rate our scientific knowledge is growing, I believe that we will one day achieve the incomprehensibly intelligent eminence of the Anunnaki, if we survive "our space age...and avoid nuclear disaster."

Although it is still to be determined how the Anunnaki came to be on their planet Nibiru, I believe that most Earthlings will one day recognize that there is only one...Creator of All.

Dr Linus Pauling's answer to my question still rings true to me..."If you wish to believe in a God as the creator, please do, as we can not explain what there was just a millisecond before the Big Bang."

ADAM

"Enki the boy child held in his hands; the image of perfection he was. He slapped the newborn on his hindparts; the newborn uttered proper sounds.
He handed the newborn to Ninmah; she held him up in her hands.
My hands have made it, victoriously she shouted.....

Shaggy like the wild ones he was not, dark black his head hair was, smooth was his skin smooth as the Anunnaki skin it was, Like dark red blood was its color, like the clay on the Abzu was its hue."

The Human genome study reported in 2001 that humans contain 25,000+ genes. SCIENCE magazine reports that the human genome contains 223 genes that do not have the required predecessors on the genomic evolutionary tree. The question is "from where did our alien genes come?"

Cuneiform Tablet

© Z. Sitchin
Reprinted with permission

WIRED january 2003

The First Cloning Superpower

Inside China's race to become the clone capital of the world.

by Charles C. Mann
photographs by Todd Eberle

MADE IN CHINA

114   **The New Cloning Superpower**
Stem-cell factories. Organ farms. Unlimited government funds. China is racing to become the clone capital of the world. **by Charles C. Mann**

REFERENCES

The Bible, Gensis the story of creation.

My research for this book has relied heavily on the books and articles authored by Zecharia Sitchin. I would suggest that anyone interested in a very detailed understanding of this subject should read any and all of "The Earth Chronicles" and articles that appear on www.Sitchin.com.

The six books in The Earth Chronicles series:

 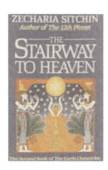 

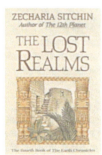 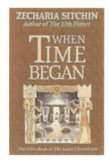 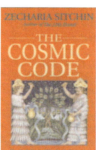

I also have researched Neil Freer's books "Breaking the Godspell", "God Games", and his outstanding October 6, 1999 White Paper, "The Alien Question".

ISBN 142519184-3

9 781425 191849